CLIFTON SUSPENSION BRIDGE

Adrian Andrews and Michael Pascoe

Clifton Suspension Bridge was my great-great-great-grandfather's first major project, which he won in open competition against the leading engineers of the day. He was only 24 years old and it must be admitted that his father helped him with the engineering details. He called the bridge *'my first love, my darling'* but sadly did not live to see it finished.

The concept was extremely daring for its time, as it was to be the longest and highest suspension bridge in the world. Unfortunately, construction was severely interrupted by the great political and social upheavals of the mid-nineteenth century. But such were Brunel's achievements and reputation that it was completed as a fitting memorial by his fellow engineers and stands today as the symbol of the city of Bristol.

Clifton Suspension Bridge receives no subsidies from local or national government but relies solely on tolls for its income. However, thanks to careful management by its trustees and the skill and dedication of a succession of Bridge Masters and maintenance staff, this majestic structure, the only surviving unaltered example of an early chain suspension bridge, should continue to be in use for many, many years to come.

On the cap of the Leigh Woods tower are inscribed the Latin words *'Suspensa vix via fit'* - a pun on the name of the original benefactor, William Vick. They loosely translate as *'A suspended way built with difficulty'*. As the readers of this fascinating guidebook will discover, rarely has a more appropriate inscription been placed on a structure.

The origin of suspension bridges

A suspension bridge is one of the oldest, strongest, most flexible and cheapest means ever devised to cross wide rivers or deep valleys.

Primitive foot bridges, with spans of up to 200m/660ft, have been built in China, India, Tibet and South America for well over two thousand years. These were made of twisted, woven vine or hide ropes, and were fast to construct and cheap to replace. The only problem was that these bridges were too flexible and could not support wheeled vehicles.

Iron chain bridges were built in China centuries before the idea reached Europe.

In 1783, Henry Cort patented a technique of adding ferric oxide to heated iron which removed most of the carbon, making the metal stronger and less brittle than cast iron. *Wrought iron* could be made fifteen times quicker this way and was therefore cheaper. This metal was very strong in tension, ideal for suspension bridges.

In 1801 the first modern suspension bridge with a level trussed road deck, hung from wrought iron chain links, was built in the USA by James Finley. The design was patented in 1809 and several bridges were built under licence. The first wire cable suspension bridge was built by White and Hazard in 1816 across the Schuylkill river in Pennsylvania.

In 1820, Captain Samuel Brown RN built the first modern suspension bridge in Britain using his patent chain links. In France iron cables were used by the Sequin Brothers at Tournon-Tain over the Rhone in 1825. The world's first long-span suspension bridge, by Thomas Telford, opened at Menai, North Wales, in 1826.

Soon, many engineers were experimenting with iron chains and cables. But many of their bridges collapsed due to structural failures caused by heavy or uneven loads, inferior ironwork or the effects of wind, which were not understood at the time.

Clifton belongs to the first generation of suspension bridges built using wrought iron bar chains. It is remarkable for its bold span and height, which were called for by its location. Yet because its completion was delayed by thirty years, it did not influence mainstream bridge development. Wire cables made of steel with deep-trussed deck structures became the preferred design method.

Clifton is unique in remaining in use practically unaltered or strengthened despite the greatly increased traffic load it supports. This is due to the quality of design, workmanship and maintenance.

Top row, left to right:
Early suspension bridges in South America.
The bridge of chains at Chukha, Bhutan, c.1783.
The first chain bridge in Britain, Co. Durham, c.174
James Finley's first bridge at Jacob's Creek, c.180
Capt. S. Brown's Patent Suspension Bridge, 1817.
Capt. S. Brown's Union Bridge, Berwick-on-Tweed
1820 - the first modern suspension bridge in Britai

Right: The principles of a suspension bridge.

How the bridge works

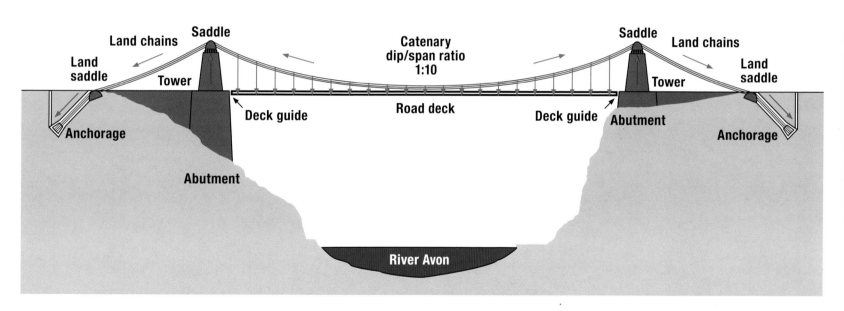

On the Clifton bridge the road deck is suspended from bar chains by iron rods. Traffic or *load* makes the rods pull down on the chains, drawing them inwards.

To resist this and to prevent the road deck collapsing, the chains are anchored into solid rock. The downward force is carried by the upward resistance of the towers.

When traffic crosses the bridge it causes the bridge parts to move. To limit the wear on the connections the weight and speed of vehicles is strictly controlled.

The parts of the bridge

Anchorages

At either end of the bridge the chains pass into an inclined tunnel 18.3m/60ft-long, cut at an angle of 45°. At the wedge-shaped base the chains are anchored, locked by iron wedges behind 1.5m/5ft x 1.8m/6ft cast iron anchor plates, weighing 3.05 tonnes/3 tons, bolted to the rock. Extra-strong Blue Staffordshire brickwork, 3m/10ft-thick, plugs the space in front of the plate.

Each anchor pit is entered by a 21m/70ft shaft covered by a metal plate in the road near the toll houses. Breathing apparatus is now worn by the maintenance crew as natural gases are emitted from the rocks.

Abutments

The two towers are built on large stone platforms called abutments. The Clifton one is low because the tower is built at the top of a steep cliff. On the other side the abutment is much higher due to the steeply sloping side of the gorge. Because no construction drawings survived, both structures were thought to be solid. But in 1978 two low-vaulted chambers were discovered in the Clifton abutment. Then, in 2002, a shaft was uncovered by the Leigh tower leading to twelve chambers 11m/36ft-high (or three double-decker buses) on two storeys, linked by very narrow tunnels (see p.31).

Towers

The two towers set 214m/702ft apart are the structures from which the chains are hung. Each tower (including cast iron cappings) is 26m/86ft-high and is estimated to weigh 4,065 tonnes/4,000 tons. They are designed to be very strong but to use the minimum amount of stone.

Brunel intended the towers to be covered by cast iron decorative panels illustrating the story of building the bridge.

From a distance the towers appear identical, but there are three differences: the corners, the shapes of the road arches and the openings on the sides of the Clifton tower. The reason for these differences is not known.

Saddles

Cars and people crossing the bridge add a 'load' which changes depending on the number and size of vehicles crossing. This causes the chains to sag slightly. The chains also expand and contract with changes in temperature. To allow for this, each chain is bolted to a 30.5 tonne/30 ton 'saddle' on top of each tower. These are mounted on rollers which allow for the slight movement caused by variations in load, and provide a strong, flexible link between the chains across the central span and the land chains. The saddles also transmit the load onto the towers.

There are also ground level saddles. These perform a similar task at the point where where the chains enter the anchorage tunnels, at either end of the bridge.

Chains

The chains are like very large bicycle chains. Each section is made up of flat, wrought iron bars that are 7.3m/24ft-long at the centre of the catenary (dip) and slightly longer closer to the saddles at the tops of the towers. This variation in length ensures that the suspension rods are set precisely 2.4m/8ft apart. The bars are connected by bolts through the holes at each end.

The design of the bars provides maximum strength with minimum weight. Computer analysis has proved that Brunel's design is close to the ideal.

Right: View of the bridge from the Clifton approach

The catenary - the dip or sag in the chains - is 21.33m/70ft. The span is 214m/702ft, giving a ratio of 1:10.

The span between the centres of the towers is 214m/702ft. The span of the road deck is 194m/636ft.

The 4,200 chain links are each 7.3m/24ft x 18cm/7in x 2.5cm/1in and weigh 0.51 tonnes/0.5 tons.

The chains are connected by 400 bolts or pins, each 63.5cm/25in x 11.7cm/4$\frac{5}{8}$ in diameter.

Cast iron capping

Saddles

Chains

Parabolic arch

Catenary

Segmental arch

Abutment

Land saddle
Anchorage

'... it must be possible to repair and even replace parts of the bridge without having to close it to traffic.'
R. F. D. Porter Goff, Brunel and the Design of the Clifton Suspension Bridge, 1974.

5

The parts of the bridge

Suspension rods

The road deck is suspended from the chains by 162 iron suspension rods. They are spaced 2.5m/8ft apart and vary in length according to their position. The rods are bolted to hangers held by the large bolts that join the links together. This simple connection forms a flexible hinged joint allowing for movement.

To equalise stress caused by loads, each rod is bolted to only one of the three chains: top, middle or bottom chain in turn, and bolted to plates on the longitudinal deck girders. Their length is adjusted by a double-threaded screw socket. They can be replaced if they break in high winds.

Road deck

The road deck is constructed from two types of iron girders. Two parallel longitudinal girders are suspended from the chains by iron rods. These 0.9m/3ft-deep wrought iron 'I' shaped girders separate the road from the footpaths. They were built in 5m/16ft sections riveted together.

Cross girders are bolted to these at right angles with diagonal ties, which are designed to make a rigid frame and brace the deck against horizontal wind pressure.

The road is made of Douglas Fir sleepers 13cm/5in-thick. Over these is a layer of planks 5cm/2in-thick, set at right angles and covered by asphalt.

Parapets

There are 1.76m/5ft 6in-wide footpaths on either side of the roadway. These were designed to enable pedestrians to enjoy the spectacular views. The parapets consist of 1.37m/4ft 9in-high wrought iron lattice-girders which provide additional rigidity to the deck structure.

These strong yet light-weight girders are fixed to elegant stancheons which are bolted to the cross girders at 2.5m/8ft intervals (matching the suspension rods).

The stainless steel strip fitted to the hardwood handrail acts as a track for the under-deck maintenance cradle, allowing it to run the full span of the bridge.

Deck guides

When the bridge was built, the engineers considered bolting each end of the road deck to the abutments. But flexibility is essential as the deck can rise and fall by as much as 25cm/10in, depending on temperature, load and wind pressure.

To allow the road deck to move up and down but to prevent it from swinging from side to side, there are two strong guides. These vital elements are located under the road deck at the junction with the abutments.

The rise and fall of the deck is best seen at the small gap between the footpath and the abutment parapet wall.

Dynamic weighbridges

When traffic crosses the bridge it causes the bridge parts to move. To limit wear and damage to the connections, the weight and speed of vehicles needs to be strictly controlled. For this reason the 2.5 tonnes/2 tons weight limit and the 40kph/25mph speed restrictions are imposed.

Weighbridges on the approach roads at either side of the bridge monitor the weight of each vehicle. If this exceeds the limit, an alarm is sounded and the toll barriers are automatically locked shut.

Right: The bridge viewed from the Portway.

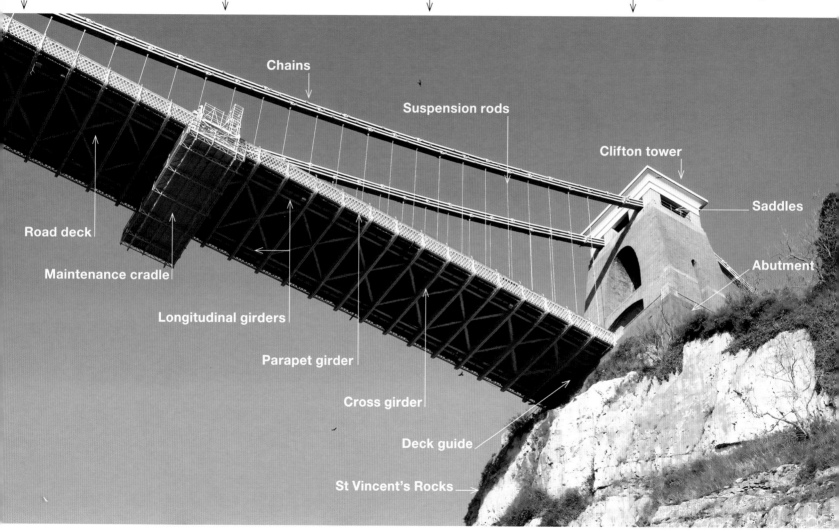

The height of the road deck above high water is 74.67m/245ft. Its overall width is 9.45m/31ft.

The two longitudinal girders weigh 113 tonnes/110 tons. The 80 cross girders weigh 81 tonnes/80 tons.

There are 162 suspension rods, from 1m/3ft to 20m/65ft-long. They weigh a total of 20 tonnes/tons.

The height of the Clifton tower is 26.21m/86ft and its estimated weight is 4,064 tonnes/4,000 tons.

Chains

Suspension rods

Clifton tower

Saddles

Road deck

Abutment

Maintenance cradle

Longitudinal girders

Parapet girder

Cross girder

Deck guide

St Vincent's Rocks

'I have adopted an apparently heavy... balustrade... to prevent the unpleasant effect of looking down from a great height.'
I. K. Brunel, letter of submission, 1830.

Bristol onion
Allium sphaerocephalon

Bristol rock-cress
Arabis scabra

Marbled white
Melanargia galathea

Peregrine falcon
Falco peregrinus

Autumn squill
Scilla autumnalis

Wilmott's whitebeam
Sorbus wilmottiana

A rare and special place

The processes that formed the Avon Gorge cannot be verified. But it is possible that when ice existed in this area, about 700,000 years ago, this was the only escape route for the meltwaters which eroded the rocks to form the gorge.

The rocks were formed about 350 million years ago as sediments on the sea bed. Sea creatures were incorporated. The rocks hardened into layers and later were tilted by earth movements to the angle we see today.

Evidence of early human occupation has been found, most significantly three large Iron Age hill forts, one on either side of Nightingale Valley and another opposite, on Observatory Hill. These defensible settlements may have once been linked by a ford.

For centuries the cliffs were quarried for their high quality limestone. The rocks also contain a quartz which was polished and sold as *'Bristol Diamonds'* to visitors to the Hotwells spa. This stood on the edge of the river below the bridge on the city side, but was demolished in 1867 and only the colonnade of former shops remains.

The river has long been the main route to and from Bristol. But its twists and turns and the unpredictable wind and currents made navigation difficult for larger sailing ships. These vessels had to be towed by *'hobblers'* (rowing boats) steered by skilful pilots. A further hazard is the tidal rise and fall of up to 9m/30ft. In 1793, a report noted that the tides made navigation of the river impossible for 150 days and dangerous for a further 89 days. The steady increase in the size of ships led to the decline of the ancient city port.

Railways were built on both sides of the river in the 1860s, to connect the old port with the new docks at the mouth of the river. The Portway was said to be the most expensive road in Britain when it was built in 1926.

As long ago as 1562, a local clergyman recorded a rare plant in the Avon Gorge. Today, the gorge is designated as a *Special Area of Conservation* and a *Site of Special Scientific Interest*. At least 35 rare plants grow here. The Bristol whitebeam and Wilmott's whitebeam are unique to the gorge and two wild flowers, the Bristol rock-cress and the Bristol onion, grow nowhere else in Britain.

The gorge is also home to some special animals. Peregrine falcons (the fastest animals in the world) nest here along with ravens, kestrels and jackdaws. Endangered lesser and greater horseshoe bats roost in the caves. The gorge is also important for insects, including the rare silky wave moth.

Roe deer have been seen swimming across the river from Leigh Woods, and there have been reports of badgers and foxes using the bridge as a night-time shortcut!

Leigh Woods comprises 178 hectares/440 acres of broadleafed ancient woodland managed by the National Trust and Natural England.

The gorge is 2.4km/1½ miles long. The River Avon is tidal with a rise and fall of up to 9m/30ft at this point.

The Clifton and Durdham Downs, comprising 178 hectares/442 acres, were secured for the benefit of the public by Act of Parliament in 1861.

The Observatory and Giant's Cave. The limestone cliffs are over 91m/300ft-high and are popular with rock climbers.

'... a wonderful passage through a mighty hill leaving on each side very high and stupendous rocks... '
'An exact delineation of the famous cittie of Bristoll...' James Millerd 1673.

Why build a bridge at Clifton?

The question is often asked by visitors but there is no simple answer.

Horse-drawn traffic rapidly increased after the coming of turnpike roads in the 1700s, causing the narrow city gateways, streets and ancient Bristol Bridge to become extremely congested.

At that time, stone and timber were the only materials available for bridge construction. The use of iron was still very much in its infancy. The Roman art of building large, concrete-cored structures had been lost and it was not until 1796 that modern concrete was produced. Even then, nothing on the scale of the gorge was attempted. Major bridges were built by military engineers or architects. The profession of civil engineer began in 1771 with the formation of the Society of Civil Engineers in London.

Another major difficulty was a condition of the Admiralty that any bridge had to be 31m/100ft above high water to allow tall-masted warships to pass beneath.

The River Avon winds 10km/6 miles from the centre of Bristol to the Severn estuary. On leaving the city it snakes through 2.4km/1½ miles of gorge. A bridge to the Admiralty's requirements nearer Bristol would have needed hairpin access roads up and down the hills to the north, as well as a huge viaduct or embankment on the south side of the river. It would have been extremely expensive and could well have required demolishing existing property and interfered with harbour traffic during its construction.

A bridge downstream of the gorge on more level land would have required massive viaducts to reach the necessary height - both labour-intensive, and costly in materials. This was demonstrated in 1974 when the M5 Avonmouth bridge was built to a height of 31m/100ft with a span of 164m/538ft. To reach this height it was necessary to construct approach viaducts which were 1,388m/4,554ft-long and the total cost was £4,200,000.

The gorge was certainly narrowest at Clifton, but also highest at 91m/300ft-above high water. The required span of 244m/800ft was far beyond a single masonry arch. (A span of 43m/140ft was achieved only after three attempts at Pontypridd in 1765.) A bridge at Clifton would therefore have needed either high abutments to narrow the span or a multi-arched viaduct on a Roman scale, both requiring vast quantities of expensive material. And yet, it was in the middle of the eighteenth century that a Bristol merchant considered Clifton the best location for a new bridge across the Avon.

Top row, left to right:
The Avon Gorge, c.1780.
Pontypridd Bridge No. 3, c.1756.
M5 Avonmouth Bridge, 1974.

Right: View of the River Avon looking towards Avonmouth and the River Severn.

Royal Portbury Dock deep-water container port, opened in 1977.

M5 Avonmouth Bridge with long approach viaducts to give height for ships to pass beneath.

River Avon snaking 10km/6 miles from the Port of Bristol to the Severn estuary.

The River Severn, with a tidal range of up to 12.2m/40ft twice daily at its mouth.

'The greatest inconveniences of Bristol are its situation, its narrow streets and the narrowness of its river... '
Daniel Defoe, 'Tour Through the Whole Island of Great Britain', 1724.

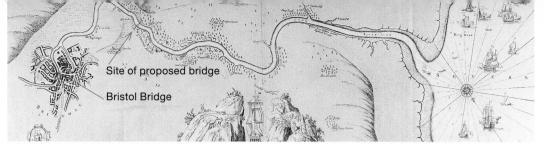

Site of proposed bridge

Bristol Bridge

1754: there was only one bridge at Bristol

'Bristol' derives from the Anglo-Saxon *'Brigstowe'* - the place of the bridge.

Until the early 1800s there was only one bridge across the River Avon at Bristol, the nearest being six miles upstream at Keynsham and no other between the city and the sea. The alternative was to cross the river by one of several small ferries, which could not take waggons or herds of cattle. Bristol Bridge was therefore the main crossing point for north-south traffic.

By 1750 Bristol Bridge was over five hundred years old. It stood on four pointed arches and was flanked by tall, narrow houses with shops beneath which projected over the river. The old bridge was 45m/150ft-long but only 4m/14ft-wide with a deep gulley in the centre of the roadway. Carts often tipped over, injuring or killing pedestrians.

Despite the demolition of the ancient city gates and the construction of a new

bridge in 1768, the problem of traffic congestion steadily worsened.

The continued charging of tolls was extremely unpopular and led to a riot. Fourteen protesters were killed or died of wounds later when the military opened fire on them.

William Vick, a wealthy wine merchant, originally from Minchinhampton in Gloucestershire, lived in the heart of the crowded city and died in 1754. In his will he left £1,000 to the Society of Merchant Venturers (the powerful guild of Bristol merchants), with the instruction that the money should be invested until it had reached £10,000. This sum he had *'heard and believed'* would be sufficient to build a toll-free stone bridge across the Avon Gorge *'from Clifton to Leigh Down'*. Vick was right in believing that a bridge would prove *'a great publick utility'*.

However, the terms of Vick's will caused *'as much amusement as surprise'*. And

rightly so, for the technology that could span such a wide and deep gorge had yet to be developed. It required a considerable leap of imagination, but William Vick's dream did become a reality over a century later.

Top row, left to right:
Detail of Captain Greenville Collins: *'The River Avon from the Severn to the Citty of Bristoll'* c.1694.
View of the old bridge.

Above: The new Bristol Bridge, 1768. Designed by James Bridges. It cost £12,00

Right: Old Bristol Bridge as depicted on James Millerd's plan of the city in 1673.

Ships sailed right up to the bridge and unloaded cargoes on both sides of the river.

Old Bristol Bridge was built in 1247, it resembled London Bridge, but on a smaller scale.

St Mary's chapel spanned the road in the middle of the bridge until 1649.

The twenty buildings on the bridge attracted some of the highest rents in the city.

Dangerous rapids formed between the bridge arches when the tide turned.

'Many limbs and lives have lost by the narrow passage of Bristol Bridge.'
Felix Farley's Bristol Journal, 17 June 1758.

1793: Bridges' bridge

Almost forty years after William Vick's death, plans for a bridge were revived. Meanwhile, much had changed. As the country's second city, Bristol's shipping and industries were flourishing. During the later 1780s a house-building boom began, especially in Clifton. The slopes of the steep hills began to be laid out for huge terraces and the richer merchants had started to move from the overcrowded city to spacious new houses on the hill.

In 1793 plans for a monumental stone bridge by the aptly-named William Bridges were published. His spectacular design would create a grand triumphal-arch entrance to the city from the sea and attract visitors to the Hotwells spa. The industries in the abutments would provide employment and the rents would pay for the structure. Although Bristol had several ironworks and trading links with Coalbrookdale, on the River Severn in Shropshire, the powerhouse of the early

Industrial Revolution, Bridges did not appear to consider using cast iron. His architectural style was also old-fashioned and the project badly timed.

Iron technology was not yet advanced. The first iron bridge with a 30m/100ft span had been built at Coalbrookdale fourteen years earlier. By 1793, the single cast iron span of 68m/220ft under construction at Sunderland would have spanned the River Avon but was far short of the 244m/800ft required to span the gorge. Engineering confidence grew swiftly. By 1800 Thomas Telford was proposing a 183m/600ft span cast iron bridge to replace the 400-year-old London Bridge.

At the same time, in Pennsylvania, USA, James Finley built the first modern suspension bridge using wrought iron chain links. Although short in span, such structures were quick to build and much cheaper than masonry bridges. Several

others soon followed as the idea spread. Within thirty years this was to provide the solution for the Clifton site.

Days after Bridges' design was published war broke out with France. The local building boom collapsed and Bristol's shipping trade was badly affected. Yet between 1804 and 1809, the city spent £600,000 on the largest engineering project in the country – the creation of the Floating Harbour. This meant that ships no longer had to settle on the mud at low tide. However, Bristol was to continue to lose trade to the cheaper, more accessible port of Liverpool.

Top row, left to right:
The first iron bridge, Coalbrookdale, c.1779.
The second iron bridge, Sunderland c.1796.
One of Finley's early chain suspension bridges, Georgetown, Washington DC. USA, c.1807.
Constructing Bridges' bridge, 1793.

Right: William Bridges' design, 1793.

A series of sketches illustrates Bridges' construction methods. One proposal shows the timber centering for the great arch being temporarily supported on suspended ropes. While a small bridge of tensioned chains existed in Britain, large-scale suspension bridges had still to be developed.

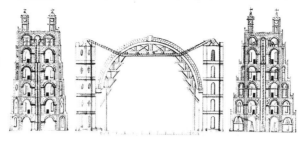

William Bridges' great arch was to be 67m/220ft high and 55m/180ft-wide. Each storey was to be 12m/40ft-high and serviced by lifts. The roadway over the arch was to be 213m/700ft-long and 15m/50ft-wide. A truly grand version of old Bristol Bridge.

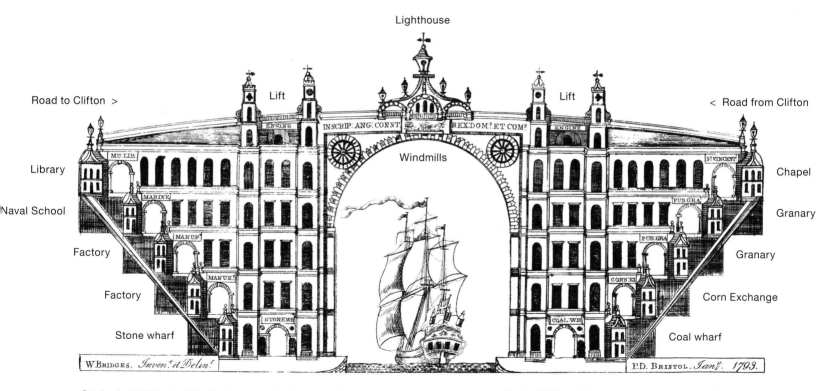

On the Leigh Woods side, factories and a stone wharf for delivery of building materials for a new suburb.

On the Clifton side, a corn exchange, granaries and a coal wharf to provide fuel for the grand new houses.

'... it has long been talked of, to build a bridge of one arch from rock to rock over the Avon.'
Matthews' New History of Bristol, or Complete Guide and Directory for the Year 1793-4.

1829: the first competition

In 1820 retired Royal Navy captain Samuel Brown built the first iron chain suspension bridge for vehicles in Europe at Berwick-on-Tweed, linking England and Scotland. His Union Bridge had a span of 138m/449ft and cost only £7,700, far cheaper than using stone. It took less than a year to build.

In Bristol, designs for a suspension bridge at Clifton were proposed in 1822 and again six years later, but came to nothing. In 1826 Thomas Telford's bridge across the Menai Straits in north Wales opened with a span of 176m/579ft. It was the largest suspension bridge yet built and attracted thousands of tourists and had a high toll income.

By 1828 the Bristol Chamber of Commerce was pressing the Society of Merchant Venturers on both the building of the bridge and the procurement of the profitable steamship mail service to Ireland. This would have required building a deep-water pier near the mouth of the River Avon on the Somerset side, to speed postal deliveries and collections by avoiding the twisting river route to and from Bristol.

In 1829 the Merchant Venturers acted, setting up a committee of commissioners and preparing to obtain the Act of Parliament required to change Vick's will, which had stipulated a toll- free stone bridge. Vick's bequest had now grown to £8,000. On 1st October the Commissioners announced a competition for an *'Iron Suspension Bridge at Clifton Down'*. Competitors had just seven weeks to design the highest and longest suspension bridge in the world.

No list of the competitors has yet been found, but various sources claim that there were twenty-two entries. These included: Samuel Brown; former Telford apprentices James Meadows Rendel and Bristol-born William Tierney Clark; Telford's trusted iron manufacturer, William Hazeldine; and his site engineer at Menai, William Rhodes.

From Birmingham, Charles Capper and Smith & Hawkes of the Eagle Foundry submitted designs. Several entries were for stone bridges and cost estimates ranged from £30,000 to almost £93,000.

Twenty-three-year-old Isambard Kingdom Brunel entered four remarkable designs with record-breaking spans. Until then he had been apprenticed to his father, the famous engineer Marc Brunel. Isambard was eager to establish a reputation for himself, but his only experience of suspension bridges had been six years previously when he worked with his father on the designs for two short-span prefabricated bridges for the island of Réunion in the Indian Ocean.

Top row, left to right:
Telford's Menai Bridge, c.1826.
Rendel's competition entry, 1829.

Right: I. K. Brunel's design no 2, 1829

16

The chains are anchored directly into the cliff, requiring no costly abutment or tower.

Two chains on each side of the road deck with short rods at the centre for stability.

299m/980ft single span, 64m/210ft above high water. Estimated cost £46,544.

Additional chains are shown under the deck to brace it against high winds.

'I thought the effect... would have formed a work perfectly unique ... the grandeur consistent with the situation.'
I. K. Brunel, letter of submission. 1830.

17

1830: Telford's cautious compromise

The committee rejected all but five of the entries on the grounds of appearance or cost, but they lacked the expertise to decide which was the best in terms of engineering. They asked 72-year-old Thomas Telford, *'the father of civil engineering'* and first President of the Institution of Civil Engineers, to advise them. Telford's Menai Bridge had been badly damaged during storms and he was therefore cautious, believing a span of 184m/600ft was the safest maximum.

Telford carefully considered each design, dismissing Brunel's plans with the words *'though pretty and ingenious would certainly tumble down in a high wind'.* Brunel's son recorded that his disappointed father withdrew his entries and *'smoked away his anger'.*

Telford's judgement was that *'none of the designs were suitable for adoption; but that Messrs. Smith and Hawkes and Mr. Hazeldine approached the nearest to practical structures'.* He recommended they should share the 100-guinea (£105) prize money but not be appointed because their schemes would require so many alterations that they would in effect be new designs. By awarding the prize money, Telford saved the committee embarrassment, but this left them with only one option – to invite the foremost civil engineer of the day to produce what they hoped would be the best possible design. Telford certainly did not reject the entries to gain the job, nor declare himself the winner as has sometimes been claimed.

Telford responded promptly and produced a design in three weeks! This was a single short-span suspension bridge, supported by two massive hollow masonry Gothic-style towers. The central span was a cautious 110m/360ft, determined by the width of the river and the positions of the two towers. He estimated the cost at £42,000, excluding approach roads, or £52,000 complete. Confident that they had achieved the best solution, the delighted committee attached the design to their Parliamentary Bill to change Vick's bequest and obtain permission to raise funds for a public work.

St. Vincent's Rocks - site of the present Clifton Suspension Bridge.

A 1.5m/5ft-wide footpath was to be located between the two carriageways, as at Menai.

The two muddy banks of the Avon defined the maximum width of the central span.

110m/360ft central span set 64m/210ft above high water. Estimated cost £42,000.

Two side spans 55m/180ft each with abutment anchorages.

'A monument of architectural taste and splendour, without parallel ... in any part of the globe.'
Clifton Bridge Prospectus, 23 January 1830.

1830: arguments and alternatives

Having travelled north to seek work, but failed, Brunel swiftly returned to Bristol and circulated a new design on pillars in the Egyptian-style costing £10,000 less than Telford's. Brunel's proposal was supported in the local press and before long there were strong criticisms of Telford's bridge. One critic claimed that the footway in the centre of the bridge, would *'preclude the enjoyment of the beautiful and picturesque scenery'*.

Opposition also came from those with vested interests, such as ferry owners and others concerned that the appearance of the gorge would be ruined. Distinguished engineer James Rendel published his design with a cheaper variation and supporting calculations, complaining that the committee had not set a budget.

There were also complaints that local talent had been ignored. Bristolians were not slow in proposing alternative designs. Coach-builder William Hill put forward two schemes. Bristol City Surveyor William Armstrong, backed by William West the artist-builder of the Clifton Observatory, re-published his design. Sign-writer William Burge proposed a monumental stone and iron-framed bridge.

In May 1830 Parliament granted permission for the bridge but only £32,000 had been raised. However, the following month, in which 24-year-old Brunel was made a Fellow of the Royal Society, King George IV died. At that time, elections followed the death of a monarch and most members of the committee were involved in electioneering.

It was not until October that the committee decided to hold a second competition, with a deadline of 18 December. They agreed to invite several previous entrants, including some local designers, to submit schemes and stated that they had no preference *'as to the adoption or non-adoption of Pillars of Suspension'*.

This time Telford entered as a competitor *'on an equal footing'*.

Top row, left to right:
William Hill's design, February 1830.
William Armstrong's design, May 1830.
William Burge's design, August 1830.

Right: Brunel's first Egyptian-style design produced in direct response to Telford's proposal, January 1830.

Below: Detail of Brunel's proposed Egyptian-style Leigh Woods tower, pen and ink drawing dated 18 December 1830.

Footpaths on both sides of the carriageway to allow visitors to enjoy the spectacular views.

Egyptian-style columns were much simpler and cheaper than Telford's Gothic-style.

110m/360ft central span 64m/210ft above high water. Estimated cost £32,000.

Two side spans 55m/180ft each with abutment anchorages.

'... the picturesque effect, and cheapness, all unite in favour of the plan of Mr Brunell (sic)*, rather than that of Mr Telford.'*
Bristol Mirror, 6 February 1830

1831: Brunel eventually triumphs

Thirteen designs were submitted for the second competition, but one late arrival was disqualified. This time the names of the entrants are known. The committee rejected all designs with pillars and short-listed entries by Brunel alongside the well-established engineers, Telford, Rendel, Brown and Messrs. Smith and Hawkes. These proposals were described as *'appearing to possess superior merit'*.

On this occasion the committee wisely decided not to consult a professional engineer to assess the designs, but rather *'one or more first rate Scientific Gentlemen'*. The person they chose was Davies Gilbert MP, a widely respected mathematician and leading theorist on suspension bridge design who had advised on the Menai Bridge. Also, Gilbert was the newly retired President of the highly influential Royal Society.

Finding a second judge proved more difficult. At the third attempt, John Seaward, a mathematician and iron manufacturer from Limehouse, London, was selected.

The designs were judged strictly on engineering qualities; the choice of the most appropriate architectural style was to be decided later by the committee. Each entry received detailed appraisal and specific criticism, but only Brunel's achieved less than the weight per square inch of chain that the judges considered necessary. However, they found fault with his proposed system of linking chains, his suspension rods and anchorages.

Telford's entry was tactfully set aside on grounds of cost and the rest placed in order of merit: 1st - Smith and Hawkes; 2nd - Brunel; 3rd - Brown; 4th - Rendel.

Brunel promptly requested a private meeting with Davies Gilbert and, after arguing with him over detailed calculations and drawings, persuaded him to reverse the decision. In his diary Brunel described this as *'persevering struggles and some manoeuvres (all fair and honest however)*. Despite this boast, Gilbert required Brunel to make several design alterations. Significantly, Telford's caution prevailed as the span was to be limited to 214m/ 703ft by the costly compromise of a larg abutment on the Leigh Woods side.

Two days later the committee endorsed the judges' decision, declaring Brunel the winner and appointing him as Projec Engineer at a fee of 2,000-guineas (£2,10 plus £500 for expenses. £850 was adde for a resident site engineer and a further £400 for an assistant site engineer.

Sadly, yet perhaps understandably, Smit and Hawkes' design has been lost. Quite what they thought of this reversal is not known.

Top row, left to right:
Davies Gilbert MP, PRS, c.1830.
The committee did not invite 16-year-ol
William Butterfield or local mason
W. W. Young to submit their designs.

Entries received from:
William Armstrong, *Bristol.*
Capt. Samuel Brown, *London.*
I. K. Brunel, *London.*
William Burge, *Bristol.*
Charles Capper, *Birmingham.*
Thomas Clarke, *Bridgwater.*
Mr. Dixon, *unknown.*
I. J. Masters, *unknown.*
James M. Rendel, *Plymouth.*
Mr. Savage, *London.*
Smith & Hawkes, *Birmingham.*
Thomas Telford, *London.*
Late entry disqualified from:
Mr. Clarke, *Hammersmith.*

William Burge - *rejected.*

William Armstrong - *rejected.*

Charles H. Capper - *rejected.*

I. K. Brunel's design no. 1 - *rejected.*

I. K. Brunel's design no 2 - *finalist placed second.*

I. K. Brunel's design no. 4 - *rejected.*

Captain S. Brown - *finalist placed third.*

James M. Rendel - *finalist placed fourth.*

Thomas Telford - *finalist but set aside.*

1831: 'the wonder of the age'

The committee's next task was to choose an architectural style. The entries had been judged on engineering detail. On 27 March, armed with sketches and models, Brunel won over the committee. In a letter he wrote: *'Of all the wonderful feats I have performed... yesterday I performed the most wonderful. I produced unanimity amongst fifteen men who were actually quarrelling about the most ticklish subject – taste... The Egyptian thing I brought down was... unanimously adopted.'*

The choice of the monumental Egyptian-style was well-judged. The massive towers were intended to contrast with the lightness of the ironwork of the chains. The towers were to be decorated with illustrations of every stage of construction. Brunel's father, Marc, took on staff and designed many of the engineering details himself.

Work began clearing, levelling and fencing off the Clifton site. On 21 June 1831 a small ceremony was held at St Vincent's Rocks. Brunel handed a symbolic stone to the wife of local landowner and investor, Sir Abraham Elton. In reply, Sir Abraham said that *'The time would come when Brunel would be recognised in the streets of every city and the cry would be raised "There goes the man who reared that stupendous work, the ornament of Bristol and the wonder of the age" '.*

This optimism was short-lived. Not only was the project £20,000 short of the necessary funds, but four months later the worst riots in 19th-century Britain erupted in Bristol. These were partly prompted by the House of Lords' rejection of the Reform Bill, which would have got rid of *'rotten boroughs'* – constituencies with few or no voters.

Widespread contempt for Bristol Corporation also added to the chaos. For three days a drunken mob ruled and ransacked parts of the city. The Custom House, Mansion House, Bishop's Palace, city jails and warehouses were looted and burnt. Brunel was enrolled as a Special Constable. A cavalry charge finally restored order, but business confidence was badly shaken and investors were reluctant to risk their money.

Work on the bridge slowed and then stopped altogether.

Top row, left to right:
Marc Isambard Brunel by E. Dubuisson.

Portrait of Isambard Kingdom Brunel, c.1835. Painting by John Calcott Horsley.

'The Bristol Riots: the burning of the Bishop's Palace', c.1831. Painting by W. J. Muller.

Leigh Woods gateway to the Clifton Suspension Bridge, 1831. Watercolour by Samuel Jackson.

Right: Brunel's winning design - a specially commissioned watercolour by local artist Samuel Jackson. April 1831.

The massive Leigh abutment built to reduce the span - a cautious and expensive compromise.

214m/702ft central span, 70m/230ft above high water, later raised to 75m/245ft. Estimated cost £52,000.

Pairs of sphinxes lie on the chains on top of the towers, facing each other across the gorge.

Two side spans 96m/315ft each half the length of the clear main span - shown here with suspension rods.

'... I anticipate a pleasant job, for the expense seems no object provided it is made grand.'
I. K. Brunel, letter to Benjamin Hawes, his brother-in-law, 27 March 1831.

1835: work starts at last

By 1835 business confidence had recovered. In August the Great Western Railway Act was passed. Brunel had already been working on this project for two years. Earlier, in February of the same year, the bridge trustees had asked him to produce a report on how the bridge might be built with funds available.

Brunel presented a simplified scheme using one pair of chains instead of two, and dispensing with the footways and gravel on the road deck. However, his design allowed for additional chains to be added at a later stage if funds allowed. In his report he wrote that it would be *by no means a contemptible work of art'* and the cost would be £35,000, just £3,000 more than the existing funds.

Optimists on the committee argued that the opening of the Great Western Railway would bring new trade and tourism to Bristol and increase commerce with Ireland and America. A deep-water pier at

Portishead would be necessary to handle larger ships and the bridge would be a vital road link to it. A more pessimistic member suggested that building a simple swing bridge over the docks entrance at Cumberland Basin would achieve the same purpose, but at far less cost. Brunel was to design just such a bridge at this location fifteen years later.

After much debate, the committee decided to keep to Brunel's original design but without the decorative iron panels on the towers. They hoped that public confidence and funds would be raised once building started. Brunel was directed to begin work again and excavations for the Leigh Woods abutment foundations commenced in December.

To attract interest in the project, a spectacular foundation ceremony was planned to take place on 27 August 1836. Subscribers were to be allotted an exclusive enclosure on the Clifton side

'being the best place for witnessing the ceremony'. The laying of the stone was to be performed by the Marquess of Northampton, President of the British Association for the Advancement of Science, which that year was due to hold its conference in Bristol.

Top row, left to right:
'Up Mail passing Reading' on the Great Western Railway. Dated 27 March 1876. Painting by B. D. Knox.

From Bristol to New York. Brunel's PS Great Western, off Portishead, embarking on her maiden voyage in 1838. Painting by Joseph Walter.

View of Cumberland Basin, the entrance to the Floating Harbour, c.1825, before Brunel's improvements. Watercolour by Samuel Jackson.

Right: A view of the proposed bridge from Cumberland Basin, Hotwells c.1836 Painting by Samuel Jackson.

One of several paintings showing the finished bridge - although building work had only just begun.

The massive Egyptian-style towers are topped with pairs of sphinxes - this time facing outwards.

The Rownham Ferry provided the only river crossing at this point - here carrying a coach and pair.

Clifton's elegant terraces of grand houses are finally complete.

'Clifton Bridge - my first child, my darling is actually going on - recommenced week last Monday - Glorious!'
I. K. Brunel, diary entry, 26 December 1835.

1836: laying the foundation stone

'Every spot on which a human being could possibly find a standing place was occupied', said a local newspaper, while ferries, boats and the new paddle steamships glided up and down the river, displaying *'all the colours they could muster'*. The event was timed at 7.30 in the morning to take advantage of the tide and to avoid interfering with the British Association's programme.

On the Somerset side, a 92m/300ft-high wooden ramp was built together with a capstan to haul up materials. Wooden stairs were built alongside the ramp, lined with flags from the ships in the harbour. Down the stairs came the procession of 400 men, walking four abreast.

A time capsule was placed in the rock containing examples of all the coins in circulation, a copy of the Act of Parliament and a specially-designed china plate from the celebratory breakfast which was to follow. An engraved copper plaque with details of the bridge covered the cavity. Using a silver trowel, the Marquess spread mortar on the plaque and the foundation stone was then lowered on top. Trumpets sounded, bands played, cannons fired salutes and the crowds cheered, clapped and waved their hats.

Three green balloons, painted as globes, were released by George Pocock and his pupils; as well as two larger 6m/20ft diameter white balloons, to one of which was attached a banner with the motto *'Success to the undertaking'*.

Three hundred ticket holders attended the breakfast and flattering speeches were made to both Brunel and his famous father, with many bouts of *'Three Cheers'*.

After the breakfast, those fit to do so visited the newly-opened Clifton Zoo, where they were serenaded by a band.

Three events slightly marred the ceremony. The symbolic first crossing of the gorge by Brunel in a basket suspended from an iron bar (which was to be used to transport workers and materials) had to be postponed due to an accident three days previously. A Captain James Swayne was drowned when his dinghy was swamped by the wash of a paddle steamer.

Lastly, the fireworks set up on the Somerset side were affected by the rain which fell during the morning. These did not perform as well as expected that evening, although a concluding set piece model of the bridge was warmly applauded.

Top row left to right:
The timber staircase and platform.

The jetty and capstan for hauling the blocks of masonry for the abutment.

One of several paddle steamers present.

Right: Laying of the foundation stone of the Clifton Suspension Bridge, 27 August 1836. Painting by Samuel Colman.

The foundation ceremony at the base of the Leigh abutment. Ships' flags line the stairs.

The four-man capstan and timber hauling way for masonry blocks off-loaded from the adjacent jetty.

A basket is shown suspended from the iron bar, although the first crossing was cancelled.

Thousands turn out to witness the scene at 7:30am. Specially chartered vessels line the river.

'... that in the course of three years the majority of this company would see this magnificent structure completed.'
Felix Farley's Bristol Journal, 3 September 1836.

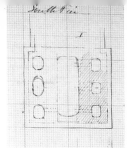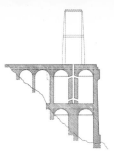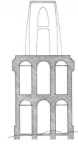

1839: slow progress

Work began on the Leigh Woods abutment in spite of insufficient funds. Within nine months and following several disputes, the contractors, Ashton and Orton, went bankrupt. The committee appointed a trustee to manage the project with Brunel *'working like the Devil'*, personally supervising building operations on site at first.

Gradually, the massive, two-storey structure took shape. The huge, vaulted chambers were built of grey Pennant rubble stone and faced with dressed ashlar blocks of Old Red Sandstone.

Letters to the local press complained of slow progress – one writer estimating that at the present rate, completion would take until 1987!

A succession of harsh winters delayed work further and the slow progress made collecting subscriptions from investors difficult, especially with the more profitable opportunities being offered by the new railways and Brunel's paddleship Great Western.

Alternative cost-saving designs were presented to the committee. In 1837 local engineer Thomas Motley proposed his *'inverted bracket suspension system'*, which was dismissed by Brunel as *'objectionable'*. The following year Brunel was asked for his opinion of James Dredge's *'patent tapering chains'*, a forerunner of modern cable-stayed bridges. Dredge had powerful supporters and the committee asked Brunel to meet them *'at a time most convenient to himself'*. But by now Brunel's services were in great demand and he was far too successful and busy to find the time.

By May 1839 work on the Clifton pier and approach road had begun, followed by the Clifton abutment. By spring 1840 the Leigh abutment was finally completed. It was 2³⁄₄ years late and had cost almost £10,000, £2,000 over estimate.

The following year both towers were completed and by 1842 the anchorage tunnels were excavated. Cranes were set up on the towers to begin construction of the chains and deck. Some of the ironwork had been delivered and all seemed set for the final phase of construction.

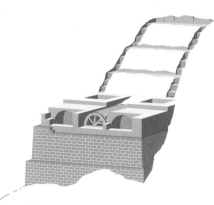

Site clearance began in December 1835 and the foundations were started in May 1836. Cornish miners were employed for the precison rock blasting.

By May 1837 the contractors went bankrupt. Brunel took over supervision and discovered a 3° error in the alignment of the foundations.

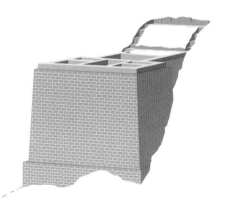

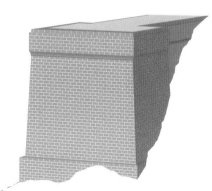

A steam engine was brought in to speed lifting operations, but construction was delayed by severe winters.

The abutment was finally completed in the spring of 1840 by two local masons - Will Williams and Philip Northam.

'... the considerable mass of masonry will require at least 12 months, probably take 15 months to build... estimated cost £8,000.'
I. K. Brunel's report to the Trustees, August 1836.

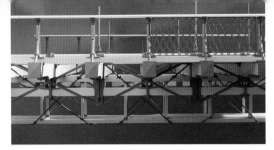

1840: precision engineering

The ironwork was the most expensive component of the bridge. Advertisements for tenders were placed in July 1839 for half the ironwork. Brunel wanted to award the contract for supplying bar iron to the Dowlais Iron Company of South Wales, with Brown Lennox of Pontypridd undertaking the forging work. But Dowlais were reluctant to bid. They thought Brunel's detailed specification and terms would make the job *'a troublesome order'*.

However, in August 1840, a contract was signed with Dowlais agreeing to supply 609 tonnes/600 tons of bar iron, to be carried by Brunel's new railway to Cardiff and then by sea to the Copperhouse Foundry at Hayle, Cornwall. Here it was forged into bar chains and the eye plates welded at each end. The Cornish foundry also cast the saddles and anchorage plates and forged the iron components for the deck.

Most of the order had been delivered to Bristol, but funds were fast running out and in 1843 Brunel instructed Copperhouse to stop production. He reported that an additional £36,348 would be required to complete the bridge.

Since selling the iron would *'destroy for ever all hopes of completion',* and despite being very busy, Brunel suggested various methods to save the project. These included a deep-water pier at Portbury, making the bridge an essential road link. The Treasury offered a loan and it was suggested that the tolls be mortgaged to the contractors, but the bridge was not considered a safe investment and none of the proposals could be underwritten.

By 1848 Copperhouse offered to accept £2,780 to settle the outstanding balance, but the offer was ignored. Copperhouse then issued a writ for £3,349. A loan of £3,250 was raised by the committee and the debt was paid off, but they refused to risk further personal liability and ordered

the sale of the ironwork to repay the loan

In January 1853 the Act of Parliament expired. The chain links were sold for use on Brunel's Royal Albert Bridge at Saltash. The anchorage pits were covered over, the cranes sold off, the Swiss Cottage site office *(see p.50)* on the Clifton side was demolished and the iron bar sold for scrap. The Clifton land was returned to the Society of Merchant Venturers, *'it being understood that the idea of completing the Bridge is now wholly abandoned'.*

Top row, left to right:
Portrait of Brunel c.1843 by J. C. Horsley.

Joseph Carne a partner in Copperhouse Foundry, a co-signatory of the 1840 contract.

A model of Brunel's deck built in 2006.

Dowlais ironworks, c.1840.
Watercolour by C. Childs

Right: Four of only nine surviving contract specification drawings - dated November 184

The lug ends were welded onto the bars with one-in-twenty bars tested to breaking point.

The deck is a forerunner of the Pratt truss. How it was to be built or would have lasted is unclear.

Forged fork ends were welded to the lower ends of the suspension rods set at 90° to the chain fixings.

The saddles were connected by struts setting the distance apart but allowing movement back and forth.

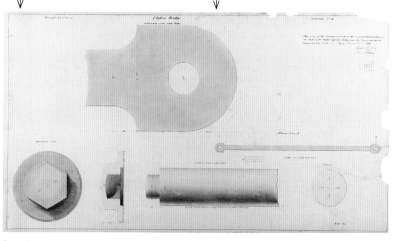

Drawing no. 2 - the chain links and pins.

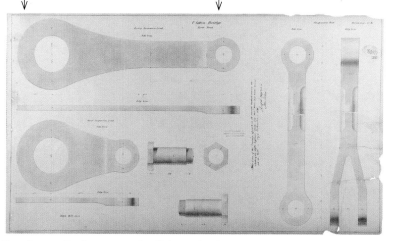

Drawing no. 3 - the suspension links and rods.

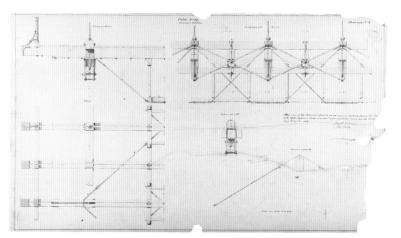

Drawing no. 7 - the floor - road deck.

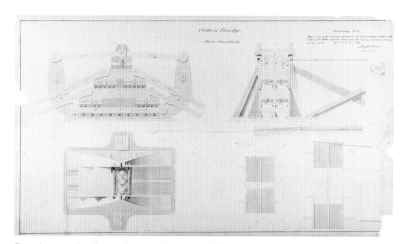

Drawing no. 5 - the main standards - saddles.

The links made at Copperhouse had *'not a difference of one-fiftieth part of an inch in the length of any two of them'.*
I. K. Brunel, during a discussion of his paper on 'Suspension Bridges' to the Institution of Civil Engineers, 1842.

1859: 'Vicksville' or 'Monuments of Failure'

There was one last attempt to rescue the project. Early in 1851, the American government's leading consultant engineer, Lieutenant-Colonel E. W. Serrell, offered to complete the bridge within fifteen months at a cost of £7,400, using lighter, iron wire cables in place of bar chains. However, the deck would be only 6m/19 ft-wide. As more details were exchanged, Serrell twice increased his estimate to £17,000. Serrell had built the Queenston-Lewiston Bridge at Niagara in 1851, linking Canada and America with a record-breaking span of 317m/1,040ft.

Still loyal to their engineer, the committee asked Brunel for his opinion. He strongly disapproved of its *'flimsiness'* but said that he would not oppose the committee. Others commented that *'the whole construction proposed was of the slightest kind'*. A new bridge company was formed and negotiations dragged on until 1857 when the proposal was shelved and Serrell was paid £90 *'for his trouble'*.

The next year, Serrell's suspension bridge at St. John's, Newfoundland, failed, followed in 1864 by his Queenston-Lewiston bridge, which collapsed as a result of extreme winter weather.

The unfinished Clifton bridge became known as *'Vicksville'*. Some Bristolians campaigned for the towers to be demolished, considering them to be *'monuments of failure'* and spoiling the scenery.

In September 1859 Brunel died, aged 53. Ironically, his death led to the completion of the bridge. The following year, John Hawkshaw, President of the Institution of Civil Engineers, was designing the Charing Cross railway bridge across the River Thames. This was to replace the Hungerford Suspension Bridge, a footbridge built by Brunel between 1841 and 1845. Hawkshaw and fellow engineer, William Barlow, wrote a feasibility study suggesting that the Hungerford ironwork

could be used at Clifton. The bar chains had been forged at Copperhouse and were the same design as those made for Clifton. The Institution of Civil Engineers also wished to finish the bridge *'as a fitting monument to their late friend and colleague',* as well as removing an embarrassment to the engineering profession.

Within a year a new company was formed and the £35,000 required to complete the bridge was raised. A new Act of Parliament was granted the following year. The Hungerford ironwork was bought for £5,000 and transported to Bristol on Brunel's Great Western Railway.

The cranes erected on both towers in 1842 for £165 had been removed in 1853 and the timber sold for only £18.

The iron bar was taken down after an effigy of a local politician had been suspended from it during an election.

The Clifton abutment had been railed off in 1853 and became a popular viewing platform for tourists and locals.

The site office was demolished and the materials sold for £5. The land was returned to the Merchant Venturers.

'I will undertake to make an arrangement... if either party refuses, the whole (matter) *must be closed as I am sick of it... '*
I. K. Brunel to G. Hennet - contractor, Private Letter Book, 6 December 1851.

1863: making the link

Hawkshaw and Barlow decided to make some major changes to Brunel's design. A pair of wrought iron longitudinal girders braced the deck, replacing Brunel's timber-trussed structure of 1840. To support the additional weight, a third chain was added. Iron parapet lattice girders contributed significantly to stiffness and strength. New anchorage pits were excavated and the land chains were shortened. An additional 203 tonnes/200 tons of chain links and 305 tonnes/300 tons of girder and ironwork were ordered. A generous donation enabled construction of the road deck to Brunel's intended width of 9.45m/31ft.

The contractors, Cochrane, Grove & Co. of Dudley, began work on site in November 1862. In April 1863, massive timber scaffolding was erected and the Hungerford saddles were installed. Next, a 2 tonne/2 ton wrought iron wire cable was fixed between the towers.

Another eight cables were then added. Stout planks were bound to six of the cables to form a walkway. Two cables above them formed hand rails and also tracks for the grooved wheels of two *'cradles'* to support the bar chains during their assembly. The final cable was at head height, from which a light-framed *'traveller'* was suspended to carry the chain links. This *'falsework'* formed a temporary bridge of great strength, but its movement was said *'to try the nerves of most persons in passing over it, oscillating as it did with every breeze'*. Despite this, construction advanced rapidly and only one fatal accident was recorded.

By June the new anchorage tunnels were ready. Anchorage plates were bolted into position and work started on assembling the chains. Twelve links were bolted to a plate, then eleven, reducing to one and two alternately up to the saddles. Each

link was hoisted to the top of the tower and carried by the traveller into position.

Working out from the saddles, a link was laid on wedge-shaped wooden blocks with the falsework giving temporary support while it was threaded on to a large bolt. Single and paired links were assembled until the chain was joined and the curve checked. The other links were then added.

Once the first chain was complete, the second was assembled above it, then the third. The process was then repeated on the northern side throughout the winter. The chains, containing 4,200 links, were finally completed in May 1864.

Top row, left to right:
Connecting the wires, April 1863.
Rigging the first 'falsework', April 1863.
Rigging the first 'falsework', May 1863.
Completion of the 'falsework' June 1863.

Right: Completion of the chains, May 1864

On 21 May 1863, a workman fell 21m/70ft, surviving for only a few minutes. The 27-year-old widower left one child.

A tunnel was being excavated for the Port to Pier Railway, intended to run from Bristol Docks to Avonmouth.

The chains were tied down by steel cables. During a storm, these broke and the falsework was thrown 21m/70ft upwards.

The chain links and saddles were brought by the GWR and hauled to site by teams of horse-drawn wagons.

'... there were some days when more than one hundred links were added... '
W. H. Barlow, Description of the Clifton Suspension Bridge, 1867.

1864: completing the job

By May 1864, work had started on constructing the road deck. As the chain links were assembled, iron hangers were set centrally from each bolt. To these were bolted the suspension rods. To equalise the load and to simplify replacement if necessary, each rod was fastened to only one of the three chains: top, middle and bottom.

The road deck was supported by two wrought iron girders which ran lengthwise, separating the road from the footpaths. These were assembled by two 5 tonne/5 ton cranes with long jibs, one on each side, which ran on narrow-gauge rails, their jibs overhanging the gorge. The cranes hoisted sections of the longitudinal girders into position, which were then bolted through plates to the suspension rods.

Beneath the girders, bolted at right angles, lattice cross girders made the structure rigid. Timber planks with rail tracks were then laid on the cross girders so that the cranes could move forward, assembling section by section. On 2 July, the last cross girder was secured and the deck structure joined.

The road deck was made of well-seasoned 12.7cm/5in creosoted Memel Baltic Pine planking, a high quality timber with a thirty-year life expectancy. This was tongue and grooved with iron. A second layer was added at right angles as the road surface. This detail avoided the need to disturb the main timbers during future repairs.

By the end of August, work was well under way on surfacing the approach roads, building the toll houses and painting the bridge. In November, the Board of Trade carried out a safety test. 508 tonnes/500 tons of stone were spread evenly over the roadway and footpaths. The deck sagged by 17cm/7in but returned to its former level when the stone was removed. The inspection was declared *'in every respec[t] perfectly satisfactory'*.

William Barlow later stated that the calculations for the form of the curve and the lengths of respective parts were mad[e] by Alfred Langley, one of his pupil assistants. Barlow proudly commented *'.. it is satisfactory to add that not a single r[od] or part of the bridge had to be altered in execution'.* A testimony indeed!

However, even by December the bridge was not completely finished. The cast iron capping on the Leigh tower had still to be fitted and left-over materials had to be hidden under the grandstand set up for the opening ceremony by the Clifton toll houses.

Top row, left to right:
Work on the deck starts, May 1864.
Constructing the deck, June 1864.
Deck girders linked, July 1864.
The completed bridge, January 1865

Right: Photograph taken from the Clifton tower, June 1864.

The photographer's assistant (circled) stands on loose planks, holding on to a suspension rod.

One of a pair of long-jib cranes used to lift each section of longitudinal girder into position.

Flag pole to provide warning signals to shipping passing below the bridge works.

The second 'falsework' supports the chains during assembly.

'... a third chain has been added, thus producing the strongest chain suspension bridge ever constructed.'
Lewis Wright, A Complete History and Description of the Cifton Suspension Bridge, February 1866.

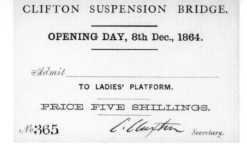
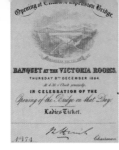

1864: Vick's dream fulfilled

The grand opening on 8 December 1864 was witnessed by many thousands of people who crowded Observatory Hill.

Members of Brunel's family did not attend. As his younger son, Henri, wrote, *'they felt that Brunel's name did not figure with sufficient prominence... the whole thing feels quite independent of any question of honouring memory'*. The changes made to Brunel's original design meant that the finished bridge was his only in form and setting.

The deck design reflected current bridge engineering developments, although by 1864 even this design was dated as wire cables were replacing bar chains and structural steel was beginning to be used.

Despite heavy rain the night before, the day was bright. To the accompaniment of pealing church bells, sixteen bands and six guns to fire salutes, a mile-long procession of societies and tradesmen carrying emblems of their businesses marched from the city centre to the bridge. Meanwhile, nobility, diplomats, Members of Parliament, magistrates, clergy and other VIPs gathered at the newly-opened Clifton Down Hotel, now Bridge House, and moved on to a grandstand through a guard of honour.

A procession then crossed the bridge under the flower-bedecked Clifton abutment. The contractors led, followed by army and navy contingents, and finally the dignitaries. After a salute had been fired, they re-crossed to Clifton where speeches were made and prayers said. To enormous cheers, the Lords Lieutenant of Somerset and Gloucestershire declared the bridge open. The great and good then adjourned to the Victoria Rooms, in Clifton, for a banquet, while the workmen were given beer and sandwiches.

In the evening, the bridge was illuminated but stormy weather had returned and *'the display failed to afford the amount of gratification to the public which had been anticipated'*.

The bridge was no longer the longest or highest in the world, but after 110 years Vick's dream had been realised and *'Brunel's bridge'* was to become the enduring symbol of Bristol.

Top row, left to right:
An unused ticket to the Ladies' Platform.

A ladies' ticket to the celebration banquet.

Engraving from the front page of the Illustrated London News.

A commemorative medal in nickel struck to mark the opening ceremony. 1200 were commissioned from a local firm, Payne & Thompson, and sold at £5 each.

Cover of the second guide book, 1866.

Right: Photograph taken from the roof of the Clifton Down Hotel, just after mid-day.

The bridge is not finished - the cast iron capping has not been fitted to the Leigh tower.

Carriages line Sion Hill, having discharged the VIPs for the opening ceremony.

The gates have just opened; the contractors and engineers lead the first crossing.

Thousands of spectators watch from Observatory Hill and Sion Hill.

'... we trust it may be seen with equal admiration by generations yet to come.'
The Times, 9 December 1864.

A unique legacy held in trust

The Clifton Bridge Company was set up in 1860 to manage the completion of the bridge. Two hundred £10 shares were allocated to the trustees in return for the land, abutments and the towers. The subsequent Clifton Suspension Bridge Act of 1861 provided that the company should pay an annual sum of £50 to the trustees to be used, with any accumulated income, to purchase debentures (bonds) and shares in the company.

For the first eighty years toll income was modest as traffic remained light, increasing only in the 1920s with the spread of ownership of private cars. Not until 1949 did the trustees acquire all the debentures and the funds with which to purchase the outstanding shares. In 1952 The Clifton Suspension Bridge Act was passed, empowering the trustees to continue to collect tolls. The revenue was to be used primarily for the maintenance and staffing of the bridge. Any surplus

was to be allocated to a 'sinking fund' to meet exceptional expenditure on major repairs, and invested for the eventual replacement of the bridge. Ultimately, if possible, it was hoped to make the bridge toll-free.

On 1st January 1953 the bridge was transferred to the trust by the company, which was then liquidated. Since then, twelve trustees, including representatives of the two local authorities linked by the bridge - Bristol City Council and North Somerset Council - together with others chosen for their engineering, technical or professional expertise, have been responsible for overseeing the running and the maintenance of the bridge.

The Clifton Suspension Bridge Trust is a non-profit-making registered charity. Its only source of revenue is from tolls, as it receives no funding or grants from central or local government or from sources such as the National Lottery Heritage Fund.

The Bridge Master, a fully qualified engineer, is responsible for the operation of the bridge. He is on call 24 hours a day and manages a staff of thirteen toll attendants and three maintenance crew.

It is easy to forget that the bridge was designed for the age of horse traffic. Today, Bristol has one of the highest levels of car ownership in Britain and over 3,000 commuters cross the bridge each weekday morning. Total vehicle crossings are more than 4,000,000 vehicles per year. To ensure this vital route remains open, the trust pursues a rigorous programme of detailed inspection and maintenance.

Top row, left to right:
The new bridge in 1865.

Troops crossing during the First World War, artillery hauled by horsepower.

Toll collecting before automation in 1975.

Right: Brunel 200 celebrations, April 2006

When the bridge was lit for special occasions, a firework display often formed a spectacular climax.

Each display seemed to surpass the previous one in terms of the creative ingenuity of the pyrotechnicians.

Huge crowds thronged the Cumberland Basin, Sion and Observatory Hills and every possible vantage point.

Today, the bridge remains a focal point of the city, and a spectacular stage set for enthralling firework displays.

'The Clifton Suspension Bridge will ever be regarded as one of the grandest conceptions of Mr. Brunel.'
The Engineer, 18 September 1864.

Meticulous maintenance

The Clifton Suspension Bridge Trust took full responsibility for the bridge in 1953. Advised by its consulting engineers, Flint & Neil Partnership, the whole structure is inspected each year, with every third inspection being more detailed. The engineers use the under-bridge cradle and abseiling techniques. Some inaccessible parts of the structure are investigated using a fibre-optic endoscope.

The wrought iron structure of the bridge has shown remarkably little corrosion, a testimony to the quality of the original manufacturer and to the standard of the engineering design and detailing. This is especially remarkable since the bridge is in an exposed position, subject to corrosive salt-bearing south-westerly winds and urban air pollution.

During its lifetime, the bridge ironwork has been protected by various paints in different colours. The chain bolts were originally picked out in gold and the chains painted with a dark grey coal tar. This was later changed to red oxide and to silver-grey in 1957. The chains were repainted in 2002 using an epoxy paint system with a pale grey acrylic top coat.

Work on the chains is undertaken from a moveable, demountable cradle. The maintenance crew have a good head for heights. They carry out painting and maintenance work on the 0.6m/2ft-wide chains at a 35 degree angle, 77 to 91m/255 to 300ft above the river, and walk up and down the chains with no hesitation. Today, safety harnesses and secure fixings are strict requirements but, up to twenty-five years ago, no safety equipment was used.

The suspension rods can be replaced and their length adjusted if they break in very high winds. This has happened on only three occasions. During one particularly violent storm, two rods on the north and three on the south side snapped. There is a special appliance which is designed to take the weight of the longitudinal girder while a suspension rod is replaced.

The programme of regular, rigorous inspection and continuous monitoring is supervised by a technical committee of specialist trustees, the Bridge Master and engineering consultants. Meticulous maintenance, combined with effective traffic management, should ensure that this remarkable bridge is preserved to provide this vital route, to and from the city, for the foreseeable future.

Top row, left to right:
An engineer inspects the chains c.1960.

Repair and replacement of a hanger bolt

Removal and replacement of a chain link for metal fatigue testing c.1970.

Right: Workmen strengthening the Leigh Woods anchorage, 1925.

In 1925, the anchorage on the Leigh Woods side was strengthened as a precautionary measure.

Here, preparations are being made to fix stays to the bridge chains prior to back-filling with concrete.

Some corrosion is evident on the underside of the upper chain, caused by poor ventilation and drainage.

Seven men are working in an extremely confined space in front of the brick skewbacks.

'... the most magnificent chain bridge ever constructed, and for strength and durability it may be pronounced unequalled.'
The Engineer, 18 September 1864.

Repairs and renewals

The annual maintenance inspections have identified elements requiring attention. Examples of major repairs include:-

1884: the upper deck timbers were replaced. This was done again in 1897 and the road was asphalted for the first time.

1925: the anchorages on the Leigh Woods side were strengthened as a precautionary measure. This was done by boring into the rock above, below and to either side of the chains, close to the upper surface of the brick skewbacks, and fixing stays to the bridge chains. The tunnel was then back-filled with 6m/20ft of concrete.

1927: the original road cradle was unbolted, lowered 75m/245ft to the road below and a new one raised.

1939: the anchorages on the Clifton side were strengthened in the same way as on the Leigh Woods side.

1955-6: the cross girders were cleaned by abrasive blasting and spray coated with zinc. The weight of zinc wire used in this process was 3 tonnes/3 tons, equal to a length of 56 Kilometres/34 miles.

1957: a new maintenance cradle was assembled from the bridge itself.

1958: all deck timbers and bolts securing the cross girders were renewed. The work took a year, during which the bridge was closed. The Clifton Toll houses were rebuilt.

1969: land slips in the gorge caused by water seepage led to an investigation into the stability of the Leigh abutment. Deep drillings around it proved the rock to be sound and the foundations safe.

1970: the bolts to sixty suspension rods were replaced and repairs made to the eyes and shackles.

1971: the pale Clipsham stone at the top of the towers was cleaned and repaired.

1972: the sway guides which prevent the deck from moving sideways were replaced with a spigot and socket system.

1978: two low-vaulted chambers were discovered under the footpaths of the Clifton abutment.

1986: the masonry of the Leigh abutment was cleaned of limescale efflorescence an accumulated bird droppings.

1998: A new maintenance cradle was hoisted into position from the river.

2002: during replacement of footpath paving, a shaft was uncovered leading to two storeys of vaults. Twelve chambers 11m/36ft-high *(equal to three double-decker buses)* were found, linked by narrow shafts and tunnels only 0.6m/2ft in diameter.

Top row, left to right:
Coating the cross girders with zinc, 195

A worker *'walks the chains'* c.1950s.

Entering the access shaft and surveying the vaults in the Leigh Woods abutment 2002.

Right: Workmen replacing the deck timbers and cross girder bolts, 1958.

In 1958-9 the deck timbers were completely replaced, together with the bolts securing the cross girders.

Approximately 70% of the heavy supporting baulks were found to be original timbers dating from 1864.

283 cubic meters/10,000 cubic feet (60 standards) of timber were used in this renewal operation.

The wooden road deck was first covered with Limmer Dock Asphalt in 1897.

'This double system of girders gives great strength and rigidity to the bridge, with very little weight of material.'
Lewis Wright, A Complete History and Description of the Cifton Suspension Bridge, February 1866.

1911

1930

1953

Lighting a landmark

On 8th December 1864, crowds flocked to see the new bridge illuminated for the first time. Under the supervision of the unfortunate Mr. Phillips of Weston-Super-Mare, the magnesium flares on the road deck were quickly extinguished by the winter winds and six electric lamps glowed only dimly in the centre of the bridge and on the towers. Limelights set around the abutments worked only now and then. The display was described as *'causing great disappointment'.*

Ornamental gas mantles were installed on the abutment parapet walls and longitudinal girders in 1865. These were replaced by electric lamps in 1928.

During the first eighty years of the 20th century the bridge was illuminated only on special occasions. In 1911 a combined display of lighting and fireworks marked the Coronation of King George V. In 1930 for Bristol-French week and in 1933, for Bristol-Brighton week, the bridge was illuminated by 1,500 electric lamps.

Some 3,000 lamps were used in 1935 for King George V's Jubilee celebrations. The bridge was lit for the Festival of Britain in 1951 and again, in June 1953, for the Coronation of Queen Elizabeth II. In 1959 the death of Brunel was commemorated and in 1964 the centenary of the bridge was marked.

Throughout the 1980s the bridge was lit from dusk to 1:00am, initially with 3,870 25w filament lamps. However, this festoon system was vulnerable to vandalism and corrosion. In addition, it presented health and safety issues when replacing lamps on the rods and chains. It was also far from energy-efficient.

A 'Guidelite' system was installed in 1992, with 27,000 30volt lights in 2.7km/1.7miles of polycarbonate tubing outlining the towers, rods and chains. Power consumption was reduced to 19kw from 95kw. But after 12 years this system suffered from dimming due to erosion of the polycarbonate, and pecking by birds.

In 2004 the Trustees engaged designers of international repute, Pinniger & Partners, to design an energy-efficient system to mark the Brunel bicentenary celebrations.

Narrow beam, low-power dimmable Light Emitting Diodes (LEDs) project light parallel to the surface of the chains. The controlled, low-intensity beams greatly reduce light-spill whilst enhancing the architectural features of the bridge.

Uplights set into the pavement are angled towards the towers, emphasising the texture of the stonework and monumental architecture, with the top of the towers capping the upward light. Precision projectors set below the deck illuminate the abutment walls and rock outcrop. These are orientated to eliminate upward light-spill *(See p.49).*

The illumination of this iconic landmark is visible for many miles. The costs of providing this widely acclaimed spectacle are met solely by the Bridge Trust.

1967

1979

1992

Tales from the Toll House

A labour of love

Before the science of geology developed, there was much speculation as to how the gorge was formed. Local legend has it that two giants, the brothers Goram and Vincent, were both in love with the fair *Avona*, who promised to marry the first man to drain the great lake that once covered the Bristol area.

Goram chose a route through Henbury Hill, whilst Vincent opted to cut through Durdham Downs. This was thirsty work so Goram stopped for a break. After a few pints he fell asleep. Meanwhile, Vincent kept digging snd completed his gorge, which drained the lake and won the girl.

On waking, Goram was broken-hearted, threw himself into the River Severn and drowned. His head and shoulders formed the islands of Flat Holm and Steep Holm.

Although unlucky in love, Goram was the star attraction at an annual local fair held at Blaise Castle, Henbury during the 1950s.

The jaws of the Japanese Juggler

A music hall artiste, Zanetto, a *'Japanese Juggler'*, attracted a large crowd when, in 1896, he is reputed to have caught a turnip which was dropped from the bridge (75m/245ft above) on to a dagger, which he clenched in his mouth!

Sarah Ann Henley, whose fashionable dress was said to have saved her life.

Skirting disaster

One bridge story continues to intrigue Bristolians. In 1885, Sarah Ann Henley, a cotton worker, jumped from the bridge after a quarrel with her railway porter boyfriend. She was wearing a fashionable crinoline and it is said that her skirts acted as a parachute so that she floated gracefully down unharmed.

A good story, but unfortunately not entirely true. Her crinoline no doubt helped to slow her fall, nevertheless she was severely injured. Her ordeal was extended by a cab driver's demand for a £10 deposit, in case her muddy dress soiled his vehicle en route to the infirmary! Lovelorn Sarah eventually recovered, married another, and died at the ripe old age of 84 in 1948.

Brunel's Swiss Cottage

Once work on the bridge had begun, a site office and store on the Clifton side became necessary. Where others might have been content with a simple hut, Brunel built a two-storey Swiss Cottage at some cost. He had earlier compared the gorge to the dramatic scenery of the Simplon Pass in Switzerland, a favourite route for wealthy tourists at this time. The office was on the first floor with windows and a balcony overlooking the site. In 1843, a shed was constructed alongside to store the ironwork. The Cottage was well maintained until the project was finally abandoned in 1853, when it was demolished and the materials sold for £5.00.

Shortly after the bridge was completed, another engineer, Francis Fox, built a Swiss Chalet-style house, *'Alpenfels'*, overlooking the gorge on the Leigh Woods side, which is still there today.

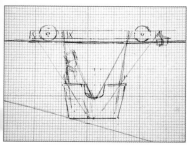

Sketch from one of I. K. Brunel's notebooks, August 1836.

The trip of a lifetime

From 1836 to 1853, the public could pay to cross the gorge in a wicker basket suspended from an iron bar. According to the *Bristol Times*, on one occasion a Somerset wedding party went to see the bridge. The bridegroom *'having had enough cider to make him adventurous, persuaded the not unwilling bride to make the flying passage'*.

The two got into the basket; but when they reached the centre of the bar, high over the Avon, it was found that the communicating rope had broken. The enthusiastic husband and his new wife dangled in mid-air for several hours. Their situation was made increasingly uncomfortable by their friends on the abutment shouting across to them that they would have to remain in the basket all night!

This novel way of starting a honeymoon was avoided but the couple were not rescued until they had spent several hours in their strange carriage while a replacement rope was obtained.

Saviour of the 'Autumn Squill'

As ground preparation work was about to start in 1831, the keen-eyed wife of site engineer, William Glennie, alerted young Brunel to the presence of a rare plant. With pioneering concern for conservation, Brunel responded swiftly, ordering the workmen to remove carefully, turf that contained the plant to a safer and less accessible location.

A small colony of *Scilla autumnalis* still survives in the gorge. A more accessible specimen may be seen in the Brunel Garden by the Clifton tower, close to its original location.

Mary Griffiths recalled how she became the first member of the public to cross the bridge in 1864.

The First Lady

The first member of the public to cross the bridge on its opening day was Mary Griffiths from Hanham, Bristol. Picking up her long skirts, 21-year-old Mary, urged on by her uncle, raced a young man from the Clifton to the Leigh Woods side and beat him by a few yards. Mary recounted her story on the radio in 1936, shortly before she died at the age of 94.

A Bristol Boxkite being flown over the bridge.

Magnificent men in their flying machines

The Clifton bridge and the gorge presented an irresistible challenge to pilots of early aeroplanes.

The first person to fly *over* the bridge was a Frenchman, Maurice Tétard, in 1910. He was demonstrating the Bristol Boxkite on behalf of the British and Colonial Aeroplane Company, formed by George White in Bristol and pioneers of aircraft building. Many pilots subsequently claimed to have flown *under* the bridge, especially during the two world wars.

In 1957, the last such flight ended tragically when a 724kph/450mph Vampire jet, after successfully flying under the bridge, crashed into the side of the gorge killing Pilot Officer John Crossley. He was marking the disbandment of No 501 (City of Bristol) Squadron of the RAuxAF.

These days, modern jet aircraft are too fast to make the attempt feasible and a legal ban on under-flying is now in force, although occasionally hot-air balloons inadvertently 'drift' under the bridge.

Acknowledgements

The authors and publishers are most grateful for the goodwill and generosity of the following in helping with the preparation of this book:

Dave Anderson, Bob Ballard, Anthony Beeson, Jane Bradley, Gerry Brooke, Mike Chrimes, Cllr. Chris Davis, Dawn Dyer, Francis Greenacre, David Greenfield, Stephen K. Jones, Mandy Leivers, Hannah Lowery, John Mitchell, Carol Morgan, Nigel Oddy, Pat Pascoe, Michael Richardson, Cllr. Howard Roberts, Mike Rowland, Alan Russell, Eileen Stonebridge and Barry Taylor.

All photographs are © the Trustees of the Clifton Suspension Bridge except for:

Adrian Andrews: *p2 top 1,2,3; p3 top 3; p10 top 1,2,3; p14 top 1,2,3; p16 top 1; p18 top 2,3,4; p22 top 1,3; p23 bottom 1,3; p24 top 1; p30 top 3,4,5; p31 illustr; p34 top 1,2,3; p38 top 4; p40 top 3,5; p42 top 1; p50 bottom 2; p51 top 2.*

Bristol Central Reference Library: *p12 top 2; p16 top 2; p23 bottom 2.*

Bristol Museums, Galleries and Archives: *p12 top 1; p13; p15 top, main; p20 top 1,2,3, bottom; p22 top 2; p23 top 1,2,3, middle 1,2,3; p24 top 2,3; p25; p26 top 1,2,3; p27; p28 top 1,2,3; p29; p30 top 1; p32 top 1; p35; p37.*

Bristol News & Media Limited: *p50 bottom 1, middle.*

BR Board (Residuary) Ltd: *p33.*

Helen Hall (AG&DWP): *p8 top 5.*

Institution of Civil Engineers, London: *p3 top 1,2; p19; p34 top 4,5.*

Chris Jones (AG&DWP): *p8 top 4.*

Rob Lisney: *p31; p46 top 3,4.*

Nick Martin (AG&DWP): *p8 top 2.*

John McAllister: *front cover.*

Morrab Library: *p32 top 2.*

National Museum of Wales: *p32 top 4.*

National Portrait Gallery, London: *p18 top 1.*

National Records Office, Kew: *p17; p21.*

Private Collection *p24 top 4.*

South West News Service: *inside front cover; p43.*

Keith Taylor (AG&DWP): *p8 top 1.*

Kelly Thomas/Butterfly Conservation: *p8 top 3.*

University of Bristol, Special Collections: *p30 top 2; p51 top 1.*

Robert M. Vogel: *p36 top 1,2,3,4; p38 top 1,3.*

Nicholas Wray (AG&WDP): *p8 top 6.*

Zed Photography: *p5; p7; p9; p49 main.*

First published March 2008.
All rights reserved.

Text © Adrian Andrews and Michael Pascoe, 2008.

Design © Adrian Andrews, 2008.

Publication in this form © Broadcast Books, 2008.

No part of this book may be reproduced by any means without the permission of Broadcast Books and the copyright holders.

Published by Broadcast Books.
Contact: Catherine Mason,
7 Exeter Buildings, Bristol, BS6 6TH
T: +44(0) 117 923 8891
E: catherine@broadcastbooks.co.uk
www.broadcastbooks.co.uk

Web orders: www.bristolpublishers.co.uk

Printed by HSW Print.
Cambrian Industrial Park. Clydach Vale.
Tonypandy. Rhondda Cynon Taff.
CF40 2XX
www.hswprint.co.uk

ISBN 978-1-874092-49-0

Right: Aerial view of the bridge and the ci